ECO-DISASTERS

NUCLEAR ACCIDENT

CHERNOBYL POWER PLANT, UKRAINE

by Meish Goldish

Consultant: Dr. Timothy Mousseau
Professor of Biological Sciences
University of South Carolina
Columbia, South Carolina

BEARPORT
PUBLISHING

New York, New York

Credits

Cover and Title Page, © Eight Photo/Shutterstock; 4, © AP Photo; 5, © Wouter van de Kamp/Shutterstock; 6L, © Miriam Doerr Martin Frommherz/Shutterstock; 6R–7, © Thegrimfandango/Dreamstime; 7TR, © SERGEY DOLZHENKO/EPA/Newscom; 8, © Ewing Galloway/Alamy; 9, © SPUTNIK/Alamy; 10, © World History Archive/Alamy; 11, The Design Lab; 12, © SPUTNIK/Alamy; 13T, © Photofusion Picture Library/Alamy; 13B, © Michał Machalski/Dreamstime; 14T, © Stefan Wisselink/tinyurl.com/lwjlxcm/CC-BY 2.0; 14B, © Alexey Wraith/Shutterstock; 15, © George Olney/Alamy; 16L, © SPUTNIK/Alamy; 16R, © EnolaBrain81/Shutterstock; 17, © Photo by Igor Kostin/Sygma via Getty Images; 18, © SPUTNIK/Alamy; 19, © SPUTNIK/Alamy; 20, © FORGET Patrick/SAGAPHOTO.COM/Alamy; 21, © AP Photo/Rainer Kostermeier; 22T, © Radiokafka/Shutterstock; 22B, © Roland Witschel/picture-alliance/dpa/AP Images; 23, © zzugu/Alamy; 24, © MarekPL/Shutterstock; 25T, © Juliane Thiere/Alamy; 25B, © AP Photo/Sergei Chuzavkov; 26, © r.classen/Shutterstock; 27, © Denis Belitsky/Shutterstock; 28, © ZUMA Press, Inc./Alamy; 29T, © PA Images/Alamy; 29B, © BrazilPhotos/Alamy; 31, © EnolaBrain81/Shutterstock.

Publisher: Kenn Goin
Editor: Jessica Rudolph
Creative Director: Spencer Brinker
Photo Research: Editorial Directions, Inc.

Library of Congress Cataloging-in-Publication Data

Names: Goldish, Meish, author.
Title: Nuclear accident : Chernobyl Power Plant, Ukraine / by Meish Goldish.
Description: New York, New York : Bearport Publishing, [2018] | Series: Eco-disasters | Includes bibliographical references and index.
Identifiers: LCCN 2017015984 (print) | LCCN 2017022748 (ebook) | ISBN 9781684022816 (Ebook) | ISBN 9781684022274 (library binding)
Subjects: LCSH: Chernobyl Nuclear Accident, Chornobyl´, Ukraine, 1986—Juvenile literature. | Nuclear power plants—Accidents—Ukraine—Chornobyl´—Juvenile literature. | Environmental disasters—Ukraine—Chornobyl´—Juvenile literature.
Classification: LCC TK1362.U38 (ebook) | LCC TK1362.U38 G65 2018 (print) | DDC 363.17/99094777—dc23
LC record available at https://lccn.loc.gov/2017015984

Copyright © 2018 Bearport Publishing Company, Inc. All rights reserved. No part of this publication may be reproduced in whole or in part, stored in any retrieval system, or transmitted in any form or by any means, electronic, mechanical, photocopying, recording, or otherwise, without written permission from the publisher.

For more information, write to Bearport Publishing Company, Inc., 45 West 21st Street, Suite 3B, New York, New York 10010. Printed in the United States of America.

10 9 8 7 6 5 4 3 2 1

Contents

A Huge Blast........................ 4
Deadly Flames 6
The Power of Radiation............. 8
A Dangerous Secret................ 10
A Smoky Battle 12
Leaving Home..................... 14
The Giant Cleanup 16
More Sickness.................... 18
Forbidden Foods 20
Angry Voices..................... 22
A Ghost Town 24
The Debate Continues 26

Fixing the Future 28
Glossary 30
Bibliography 31
Read More 31
Learn More Online 31
Index 32
About the Author 32

A Huge Blast

The early morning of April 26, 1986, was calm outside the city of Pripyat, Ukraine. Then, around 1:30 am, everything changed. Workers at the Chernobyl **Nuclear** Power Plant were testing equipment that controlled the plant's Unit 4 **reactor**. Suddenly, the plant started to shake violently. A power increase caused a huge steam explosion inside the reactor. The blast blew off the reactor's 1,100-ton (998 metric ton) **concrete** roof! Two workers died immediately in the explosion.

Nuclear power plants produce electricity for businesses and houses.

This picture shows the Unit 4 reactor after the explosion. The power plant had four units, each with a reactor.

The tremendous heat from the blast caused more than 30 fires to break out in the Unit 4 reactor. Giant flames lit up the night sky. People in Pripyat could see the fires from their homes. One **resident** said, "The whole sky [was] a tall flame. . . . The heat was awful."

The city of Pripyat, Ukraine, was built to house the plant workers and their families.

The Chernobyl Nuclear Power Plant is about 2 miles (3.2 km) outside the city of Pripyat.

Deadly Flames

More than 100 firefighters from Pripyat raced to the power plant to battle the fires. They sprayed water on the flames but were unable to put them out. Unlike ordinary fires, these nuclear fires were burning extremely hot because they were being feuled by a material called **graphite**. One firefighter remembered, "We tried to beat down the flames. We kicked at the burning graphite with our feet." Yet, nothing worked.

Graphite is used in nuclear reactors to control **chain reactions** that create heat. The heat makes steam, which turns large machines to create electricity.

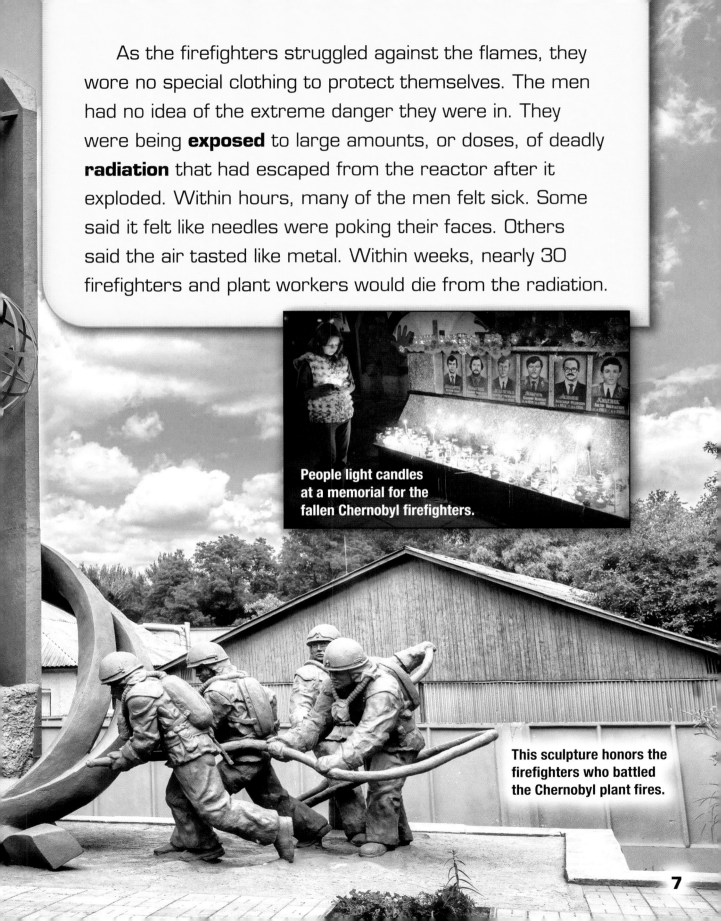

As the firefighters struggled against the flames, they wore no special clothing to protect themselves. The men had no idea of the extreme danger they were in. They were being **exposed** to large amounts, or doses, of deadly **radiation** that had escaped from the reactor after it exploded. Within hours, many of the men felt sick. Some said it felt like needles were poking their faces. Others said the air tasted like metal. Within weeks, nearly 30 firefighters and plant workers would die from the radiation.

People light candles at a memorial for the fallen Chernobyl firefighters.

This sculpture honors the firefighters who battled the Chernobyl plant fires.

The Power of Radiation

Why is radiation in large amounts so dangerous? It attacks body cells that keep a person healthy. Often, victims of radiation sickness may vomit and get burns on their skin. People who breathe radiation can get lung damage. Over time, they may also develop **cancer**. That's what happened to many firefighters and plant workers after the accident at Chernobyl.

The Chernobyl explosion released 400 times more nuclear **fallout** than the atomic bomb dropped on Hiroshima, Japan, during World War II in 1945.

Large amounts of radiation are powerful enough to spread from an object to a person. One Chernobyl worker said, "I took off all the clothes that I'd worn there and threw them down the trash chute. I gave my cap to my little son. He really wanted it. And he wore it all the time. Two years later, [he had] a **tumor** in his brain."

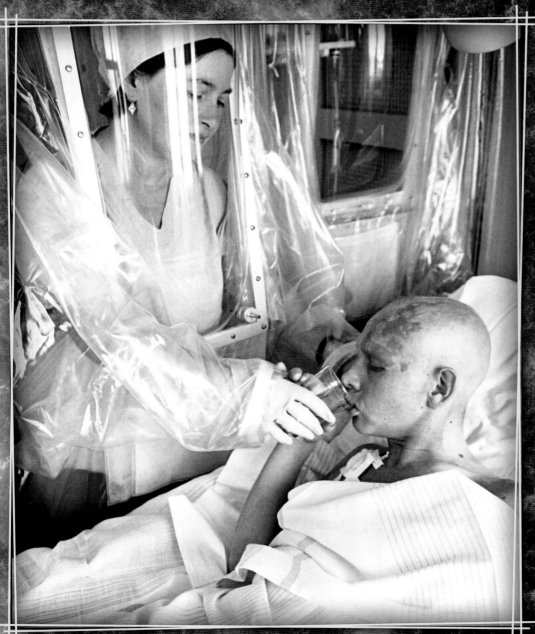

Some doctors and nurses refused to treat Chernobyl radiation victims, fearing they would get sick themselves.

A Dangerous Secret

Ukrainian officials knew that the Chernobyl plant accident was extremely serious. Even so, they tried to keep it a secret. Ukraine was part of the Union of Soviet Socialist **Republics** (USSR). The USSR was very proud of its nuclear program and did not want to admit that any failure had occurred. Soviet leaders hoped to keep other countries—and even their own people—from learning about the disaster.

Mikhail Gorbachev was the leader of the Soviet Union at the time of the Chernobyl accident.

The Union of Soviet Socialist Republics was a large country made up of 15 republics, all under **Communist** rule. The USSR existed from 1922 until 1991.

However, other nations soon discovered the secret. Two days after the blast, scientists in Sweden and other European countries **detected** high levels of radiation in the air and on the ground. High winds had carried it from Ukraine. With the Chernobyl plant still burning, Soviet officials could no longer hide what had happened. They publicly admitted to the disaster.

On April 28, scientists in Sweden, Finland, Denmark, and Norway found high levels of radiation. Sweden is 722 miles (1,162 km) from Chernobyl.

A Smoky Battle

For more than a week, the fires inside the Unit 4 reactor continued to burn. Firefighters in helicopters dropped sand and clay into the reactor to **smother** the flames. They also dumped lead and boron. They hoped these two substances would **absorb** the radiation and help remove it from the air and ground.

A helicopter flies over the Unit 4 reactor at Chernobyl.

The plan was only partly successful. As they flew over the reactor, firefighters were not able to see clearly through the thick smoke. They often missed their target below. Even worse, many firefighters became sick and some even died soon after flying into the radiation. Yet, finally, after continuing for ten days, the fires at Chernobyl burned themselves out.

The gas masks, helicopters, and other equipment used to fight the Chernobyl fires had to be abandoned because they had become **radioactive**.

Leaving Home

The nuclear accident had a lasting effect on the city of Pripyat. Radiation from the Chernobyl plant had **contaminated** the entire city. The day after the blast, officials told the 48,000 residents that they would have to leave their homes for three days while the town was cleaned. The city used 1,100 buses to move all the residents to tents in the woods until the cleanup was finished.

After the people of Pripyat were forced to leave, all areas of the city, including apartment buildings (above) and this amusement park (right), were suddenly empty.

The people of Pripyat had mixed feelings about the forced **evacuation**. Lyudmilla Ignatenko, who was married to a firefighter, recalled, "The whole street was covered in white [cleaning] foam. We were walking on it, just cursing and crying. Some people were glad—a camping trip! Only the women whose husbands had been at the reactor were crying." Two weeks later, Lyudmilla's husband died of radiation sickness.

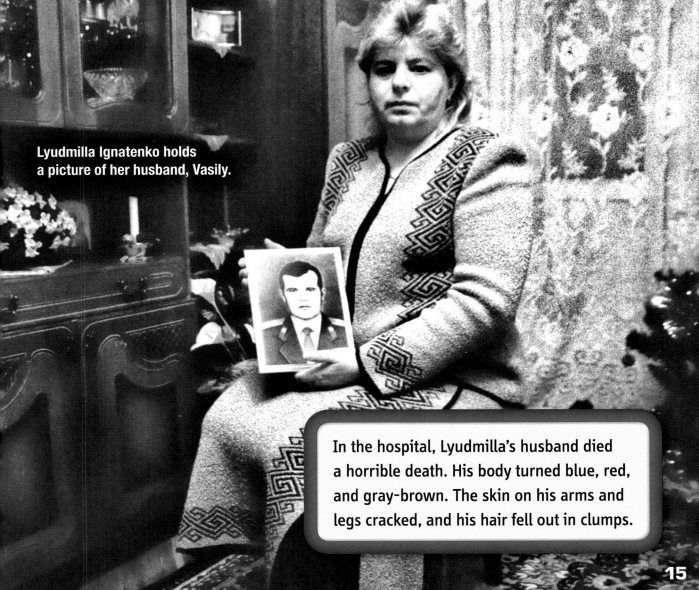

Lyudmilla Ignatenko holds a picture of her husband, Vasily.

In the hospital, Lyudmilla's husband died a horrible death. His body turned blue, red, and gray-brown. The skin on his arms and legs cracked, and his hair fell out in clumps.

The Giant Cleanup

With Pripyat now empty, a cleanup of the city began. Workers called liquidators scrubbed the streets and buildings with water and chemicals. However, scientists soon realized that radiation had contaminated other towns beyond Pripyat. In May, officials set up an area called an "**exclusion** zone." All people living within 18.6 miles (30 km) of the nuclear plant—about 135,000 people—had to leave their homes and move to other cities.

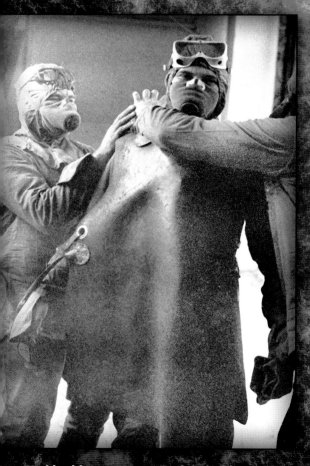

Liquidators wore special clothing to help protect them from radiation as they washed buildings.

Children had to leave behind dolls and other radioactive toys when they left their homes.

The people who were forced to leave the Chernobyl Exclusion Zone were not allowed to return. Buildings, plants, and water in the zone were radioactive, making it an unhealthy place to live.

Soon, about 600,000 liquidators were working in the zone. After the power plant fires went out, workers could wash the reactor and other equipment in the Unit 4 reactor. Some liquidators had the job of building a giant concrete shell, called the **Sarcophagus**, over the broken reactor. The huge covering prevented more radioactive materials from escaping into the air.

Chernobyl liquidators clear radioactive litter from a reactor rooftop.

More Sickness

The zone cleanup continued for several months, and many liquidators developed radiation sickness. Some became ill while they worked. Others did not start to get sick until years later. Over time, many liquidators' children came down with cancer or other diseases. Some of the children who became sick were born after their fathers stopped working at Chernobyl.

Scientists **estimate** that anywhere from 4,000 to 200,000 people who were near the Chernobyl plant after the explosion will eventually die of cancer.

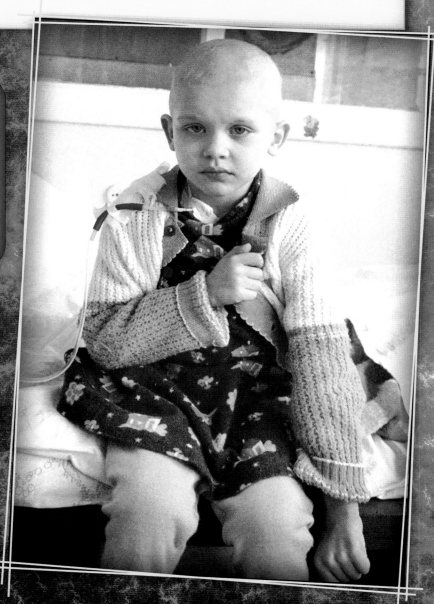

A child with radiation sickness is treated in a hospital.

Many people who moved from the Chernobyl Exclusion Zone to other towns suffered even if they weren't ill. One mother said, "When we settled in Mogilev, and our son started school, he came back the very first day in tears. They put him next to a girl who said she didn't want to sit with him because she thought he was radioactive. The other kids were afraid of him."

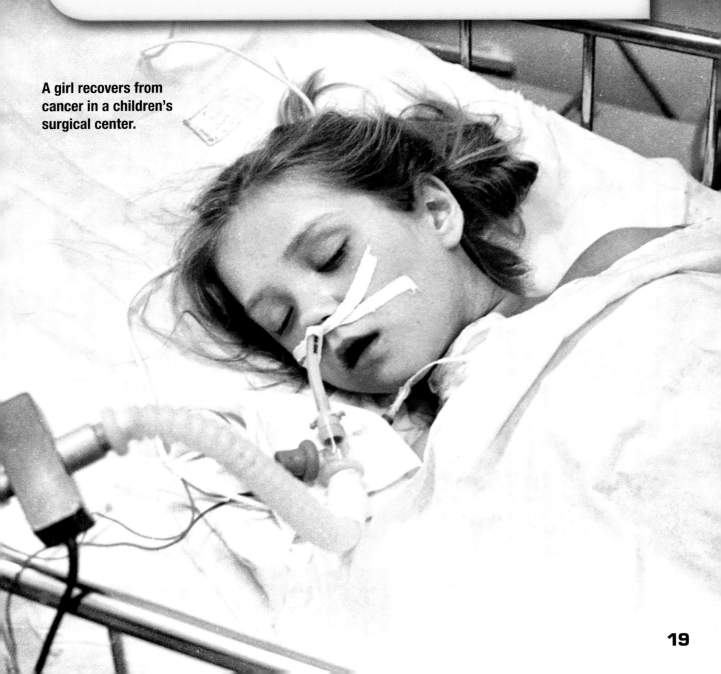

A girl recovers from cancer in a children's surgical center.

Forbidden Foods

For some people who became sick, the cause of their illness was eating and drinking contaminated foods and liquids just before they moved out of the Chernobyl Exclusion Zone. The plants and animals in the zone had been exposed to large amounts of radiation. Many children developed cancer after drinking milk from radioactive cows.

Over time, many animals in the Chernobyl Exclusion Zone became sick, stopped **reproducing**, and died.

Winds blew the radiation from Chernobyl across the rest of the USSR, Europe, and as far as the United States and Canada. In many countries, officials took steps to prevent illness and death. In Austria and England, farmers were not allowed to sell foods from areas that had been contaminated with radiation. In Canada, people were told not to drink rainwater. In Poland, medical workers gave special tablets to children to prevent them from developing **thyroid** cancer.

A worker collects contaminated vegetables so they can be destroyed.

Angry Voices

Soon after the Chernobyl disaster, people in many countries began protesting against nuclear power plants. They feared that similar accidents could occur elsewhere. Protestors demanded that world leaders cancel the openings of nuclear plants that were under construction, and shut down those that already existed.

A protestor holds signs with a clear message.

People march to protest the use of nuclear power.

Soviet officials considered closing the Chernobyl plant for good. However, with a bitter, cold winter ahead, the nuclear plant would be needed for its electricity. Without it, many people might freeze to death. After the Chernobyl power plant was cleaned, the reactors in Units 1, 2, and 3 were put back in operation by the end of 1986.

Before the accident, workers at Chernobyl had begun to build two new reactors—Units 5 and 6. However, construction stopped after the accident when protestors objected.

By 2000, all four reactors at Chernobyl were shut down permanently due to smaller accidents that occurred at the plant after 1986.

A Ghost Town

After the disaster, the city of Pripyat became a ghost town—and remains one today. No one is allowed to live there because of the deadly radiation that still lingers. The catastrophe has left many former residents very **depressed** because of the risk of illness. One said, "We're afraid of everything. We're afraid for our children, and our grandchildren, who don't exist yet."

Experts believe the Chernobyl Exclusion Zone won't be safe for people to live in for another 20,000 years!

Although living in the Chernobyl Exclusion Zone is illegal, about 200 former residents have snuck back in. They found it too difficult to make new lives for themselves in other places. They live in contaminated houses, despite the danger. One resident said, "Even if it's poisoned with radiation, it's still my home. There's no place else they need us."

Maria Ilchenko is an illegal resident in the Chernobyl Exclusion Zone.

Many wild animals, such as foxes and wolves, have moved into the Chernobyl Exclusion Zone. Scientists are not sure how the radiation will affect them in the future.

The Debate Continues

The 1986 Chernobyl disaster was the world's worst nuclear accident. It created a **debate** about nuclear power that continues today. Some people believe that all nuclear power plants should be shut down because they are too dangerous. They say other sources of energy, such as clean energy and even coal, oil, and gas, should be used instead.

There are currently hundreds of nuclear reactors operating in dozens of countries around the world.

Clean energy uses **renewable resources**—wind, sunlight, and flowing water. Using these resources creates less pollution than coal, oil, or gas.

Other people disagree. They argue that burning coal, oil, and gas creates too much air pollution. Nuclear energy doesn't pollute, as long as the radiation is controlled. Nuclear power also provides more energy than many other sources. As the world's population continues to grow and demand more energy, relying on clean energy and nuclear power plants—that are run with high safety standards—may be a practical answer.

This coal-powered factory in Ukraine is creating air pollution.

Fixing the Future

Since the Chernobyl nuclear disaster, actions have been taken to try to prevent other nuclear disasters from occurring in the future. Here are some examples.

The Next 100 Years

A new sarcophagus, called the New Safe Confinement facility (NSC), is being built at the Chernobyl nuclear power plant. It's scheduled to be finished by November 2017. The new shelter will provide a stronger cover around the Unit 4 reactor than the previous sarcophagus. It will prevent radiation from leaking into the air for the next 100 years.

The new sarcophagus around Chernobyl's Unit 4 reactor

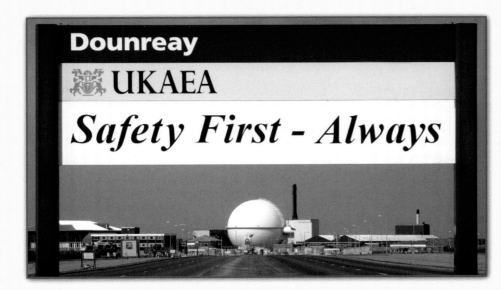

A sign at a nuclear power plant in the United Kingdom

Stronger Shells

Today, reactors at other nuclear plants have stronger structures built around them than the ones at Chernobyl did. The new protective shells prevent radiation from escaping, even in the event of an earthquake.

Open Information

New rules have been made to ensure that nuclear accidents cannot be kept secret. In the United States, if an accident occurs at a nuclear plant, it must be announced within 15 minutes so people can evacuate immediately.

Practice Drills

Practice drills are held at various times to test how quickly and effectively plant workers can respond to an emergency. Every two years in the United States, drills are observed by government officials.

Nuclear power plant workers wear protective suits when handling dangerous materials.

Glossary

absorb (ab-ZORB) to soak up something

cancer (KAN-sur) a serious, often deadly disease that destroys parts of the body

chain reactions (CHAYN ree-AK-shuhnz) nuclear reactions producing energy that cause further reactions

Communist (KAHM-yuh-nist) relating to a system of government that limits the amount of property people own

concrete (kahn-KREET) a mixture of sand, water, cement, and gravel that is used in construction

contaminated (kuhn-TAM-uh-nay-tid) made polluted or unfit for use

debate (dih-BAYT) a discussion of a problem or issue

depressed (dih-PREST) very sad

detected (dih-TEK-tid) discovered or noticed something

estimate (ESS-tih-mayt) to make a careful guess about the size, cost, or value of something

evacuation (ih-vak-yoo-AY-shuhn) the removal of people from a dangerous place

exclusion (ek-SKLOO-zhuhn) the act of leaving something or someone out

exposed (ek-SPOHZD) made open to danger such as a disease

fallout (FAWL-out) the radioactive particles that are stirred up from a nuclear explosion and descend through the air

graphite (GRAF-ite) a soft black or gray form of the element carbon that slows down radioactive particles

nuclear (NOO-klee-ur) having to do with a type of energy that is produced by splitting atoms

radiation (ray-dee-AY-shuhn) a form of energy that can be very dangerous when not properly controlled

radioactive (ray-dee-oh-AK-tiv) giving off dangerous, invisible rays of energy

reactor (ree-AK-tur) a machine in which uranium atoms are split to produce nuclear energy

renewable resources (re-NOO-uh-buhl REE-sorss-iz) resources such as water, wind, and sunlight that are continuously renewed or replaced by natural processes

reproducing (ree-pruh-DOO-sing) having offspring

republics (rih-PUB-liks) countries in which people vote for representatives who manage the government

resident (REZ-uh-duhnt) a person who lives in a certain place

sarcophagus (sahr-KAHF-uh-guhss) a stone box built to contain a coffin

smother (SMUTH-ur) to cover something thickly so air cannot reach it

thyroid (THYE-royd) a gland in the neck that releases hormones, which control body growth and development

tumor (TOO-mur) an unusual lump or growth in the body

Bibliography

Alexievich, Svetlana. *Voices from Chernobyl: The Oral History of a Nuclear Disaster.* New York: Picador (2005).

Mittica, Pierpaolo. *Chernobyl: The Hidden Legacy.* London: Trolley Books (2007).

Petryna, Adriana. *Life Exposed: Biological Citizens After Chernobyl.* Princeton, NJ: Princeton University Press (2013).

Read More

Benoit, Peter. *Nuclear Meltdowns (A True Book).* New York: Scholastic (2012).

Bryan, Nichol. *Chernobyl: Nuclear Disaster.* Milwaukee, WI: World Almanac Library (2004).

Johnson, Rebecca L. *Chernobyl's Wild Kingdom: Life in the Dead Zone.* Minneapolis, MN: Twenty-First Century Books (2015).

Rissman, Rebecca. *The Chernobyl Disaster (History's Greatest Disasters).* Minneapolis, MN: ABDO (2014).

Learn More Online

To learn more about the Chernobyl nuclear disaster, visit
www.bearportpublishing.com/EcoDisasters

Index

air pollution 27
animals 20, 25
Austria 21

Canada 21
cancer 8–9, 18–19, 20–21
Chernobyl Exclusion Zone 16–17, 19, 20, 24–25
clean energy 26–27
coal 26–27

England 21
Europe 11, 21
evacuation 15

firefighters 6–7, 8, 12–13, 15
food 20–21

gas 26–27
graphite 6

helicopters 12–13
Hiroshima, Japan 8

Ignatenko, Lyudmilla 15
Ignatenko, Vasily 15
Ilchenko, Maria 25

liquidators 16–17, 18
Mogilev, Belarus 19

nuclear power 4–5, 6, 10, 22–23, 26–27, 28–29

oil 26–27

plants 16, 20–21
Poland 21
Pripyat 4–5, 6, 14–15, 16, 24
protestors 22–23

radiation sickness 7, 8–9, 15, 18
radioactive 13, 16–17, 19
reactors 4–5, 12–13, 17, 23, 28–29

safety measures 28–29
Sarcophagus 17, 28
Sweden 11

Ukraine 4–5, 10–11
United States 21, 29
USSR 10–11, 21

water 6, 16, 21

About the Author

Meish Goldish has written more than 300 books for children. His book *City Firefighters* won a Teachers' Choice Award in 2015. He lives in Brooklyn, New York.